SANSKRIT LINGUISTICS AND GOD

THE LOST LINK BETWEEN MODERN SCIENCE AND GOD

NISHANTH MEHANATHAN

I dedicate this book to my family, friends, teachers, *Shree Krishna*, *Shani Bhagawan* and *Vaishnodevi Ma*

Contents

Acknowledgements

I would like to thank all my teachers, friends, colleagues and family members for the warm support they have extended to me while working on this book, these were the giants on whose shoulders I stood.

The search for God has been a tale spanning 2 decades. It has been another journey in itself, it kept me alive through these years. I would like to thank all the people who I came across on this journey for their kindness, ideas, clues and support for that is what made this work possible. I would like to thank ISCKON for their wonderful Prasdams, Darshans and Kirtans. Also Srila Prabhupada for keeping the spirit of the journey alive.

I would like to thank God who helped me in this journey to find him and gave me this opportunity to present him to the world.

Introduction

In this book, we discuss Sanskrit linguistics, and how it provides the lost link between Modern Science and God.

The technical topics covered to develop the theory required to link Modern Science and God:

What Is An Object?
Classification Of Words In the Sanskrit Language
"*Satta*" Or "existence/being" is The Highest Universal

The study of these technical topics surprisingly leads to God. The missing link between modern science and God has been found. And this work presents that important link and the identity of God along with his scientifically provable attributes. Modern science can finally prove the existence of God and his attributes.

CHAPTER ONE

A Thing

An object is defined as that which is the substratum of the genus, differentia and action [1] as depicted in Figure 1.

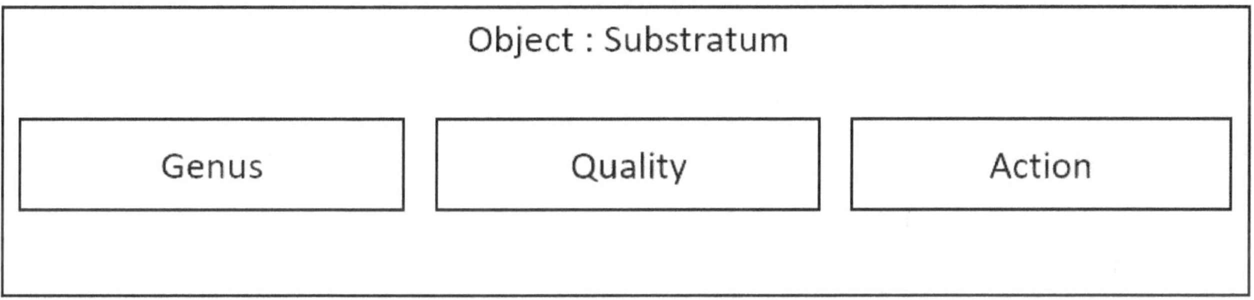

Figure 1. Object as the substratum of genus, differentia and action

These attributes (Genus, Quality and Action) are classified as shown in Figure 2. An attribute is of two kinds- one that is inherent and the other which is imposed upon it like a name. The inherent attribute again is of two kinds, an attribute that is fully accomplished and that which is in the process of accomplishment. An accomplished attribute is of two kinds- the genus or class and the quality. Genus is never found away from the individuals in which it resides, while a quality distinguishes a thing from other things belonging to the same genus or class.

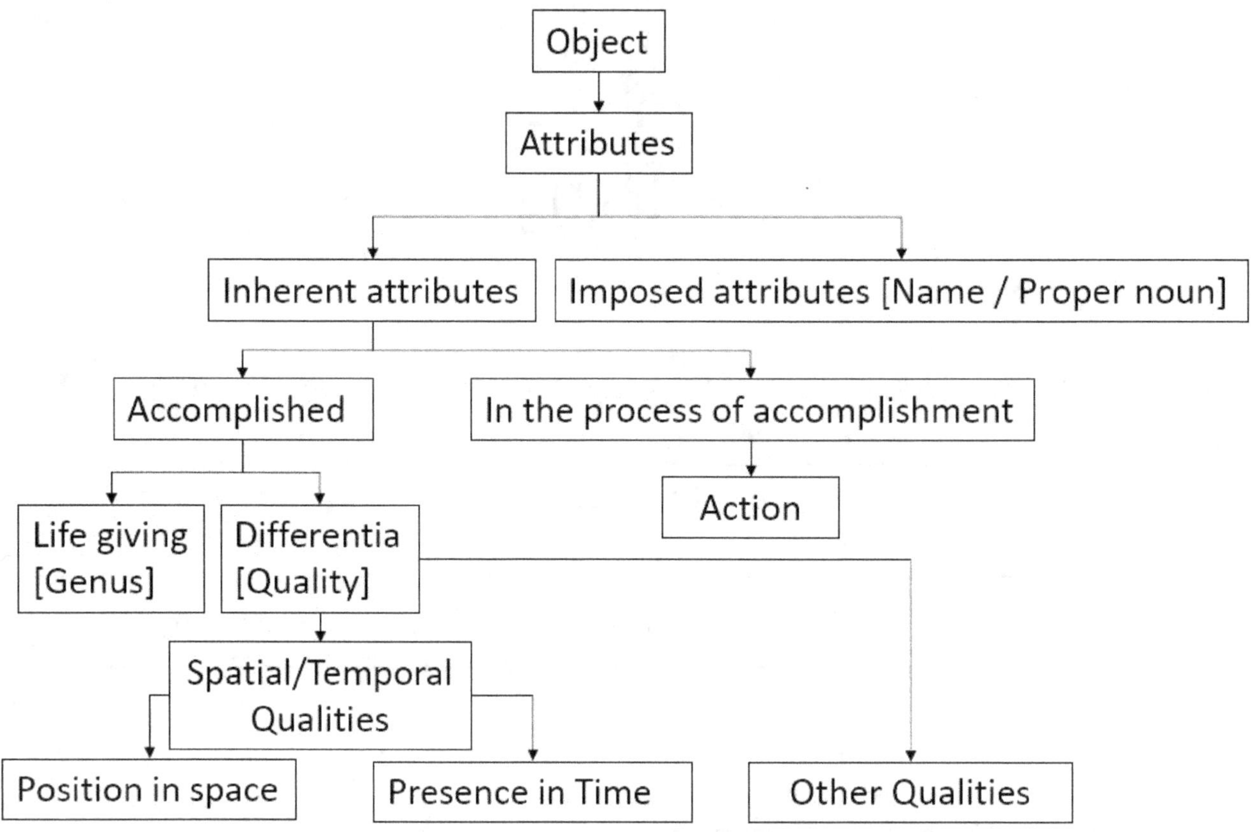

Figure 2. Object attributes, which uniquely identify an object

An example of the genus is fox-ness, quality is whiteness and action is cooking. An attribute of an object in the process of completion is called an action. For instance, the colour of a sheet of paper (whiteness) is an accomplished fact. But action implies a series of activities some completed and some in the process of completion, which occupies successive portions of time.[2].

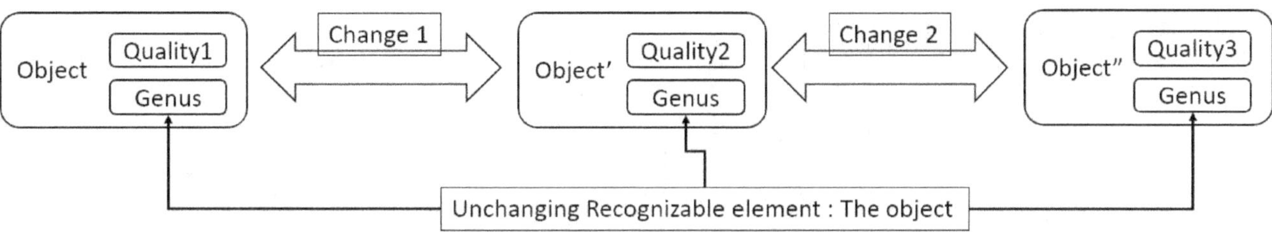

Figure 3. The unchanging recognizable element which remains across various changes

Object, its identity and the identity theorem:
Identity theorem: An object stands for that unchanging recognizable element which persists in all the changes that it undergoes [3] as shown in Figure 3. This is because if an object changes, it still remains the object hence the object is something which persists across the change and which we know to be the genus or universal.

For example, if a silver bell (or a silver object) is twisted out of shape what remains is it being made of silver and it is an object. So it is still a silver object but the silver bell is destroyed. So speaking practically it is the genus which the object is. Qualities are properties which undergo a change. As a result of the genus persisting you are able to identify the object across all changes as the genus remains constant.

Classification of words in the Sanskrit Language and the types of objects a word can refer to:
The Sanskrit linguists hold that the import of words is either genus, quality, action or names. Hence there are four main types of words- genus word, quality word, action word and name word [4]. It is particular to the Sanskrit language but the import or idea is universal to all languages. What is indicated by a word is of four types- genus, quality, action and proper names as depicted in Figure 4.

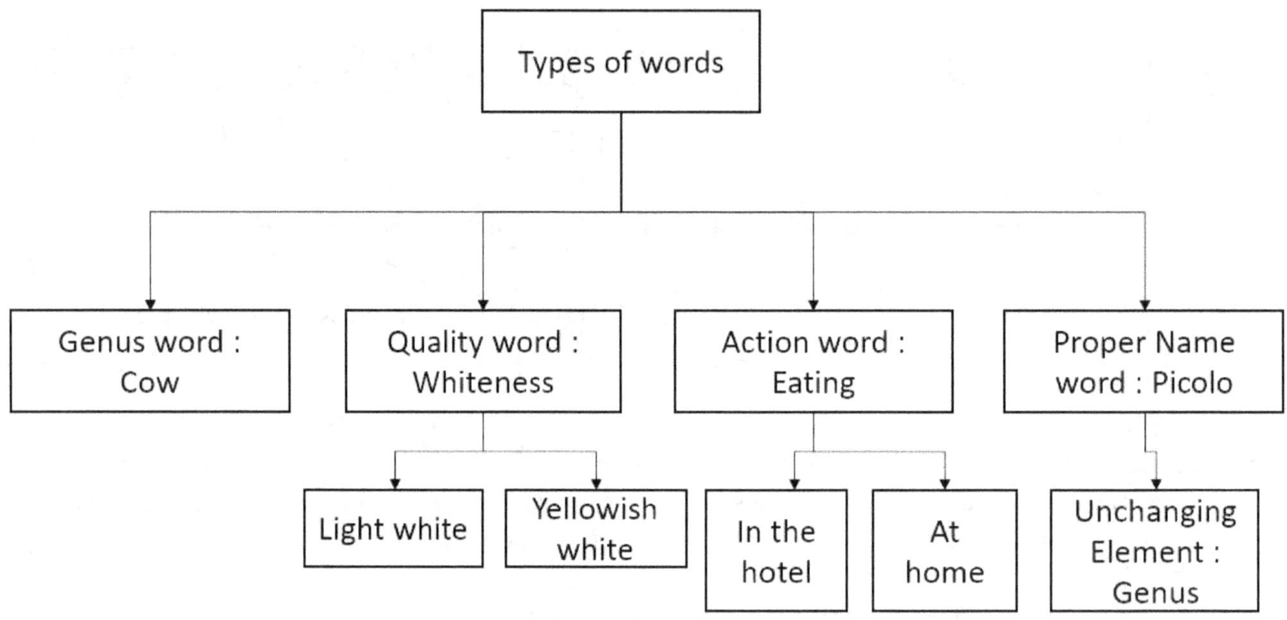

Figure 4. Types of words and what they indicate with examples.

Notes

1. Charudev Shastri, Vyakaran Mahabhashya (Pratham Aahrikantri) – Page 3.
2. Sahitya Darpan, II.4, page 43.
3. Vakyapadiyam of Bhartrhari, III.1.11, III.1.12 commentary.
4. Makkhanalāla Śarmā, Bhāratīya kāvyaśāstra ke siddhānta - Page 69.

CHAPTER TWO

All Words Convey "Existence / Being"

As discussed a thing has four attributes:

1. Name
2. Class
3. Quality
4. Action

A thing and its various attributes is depicted in Figure 1.

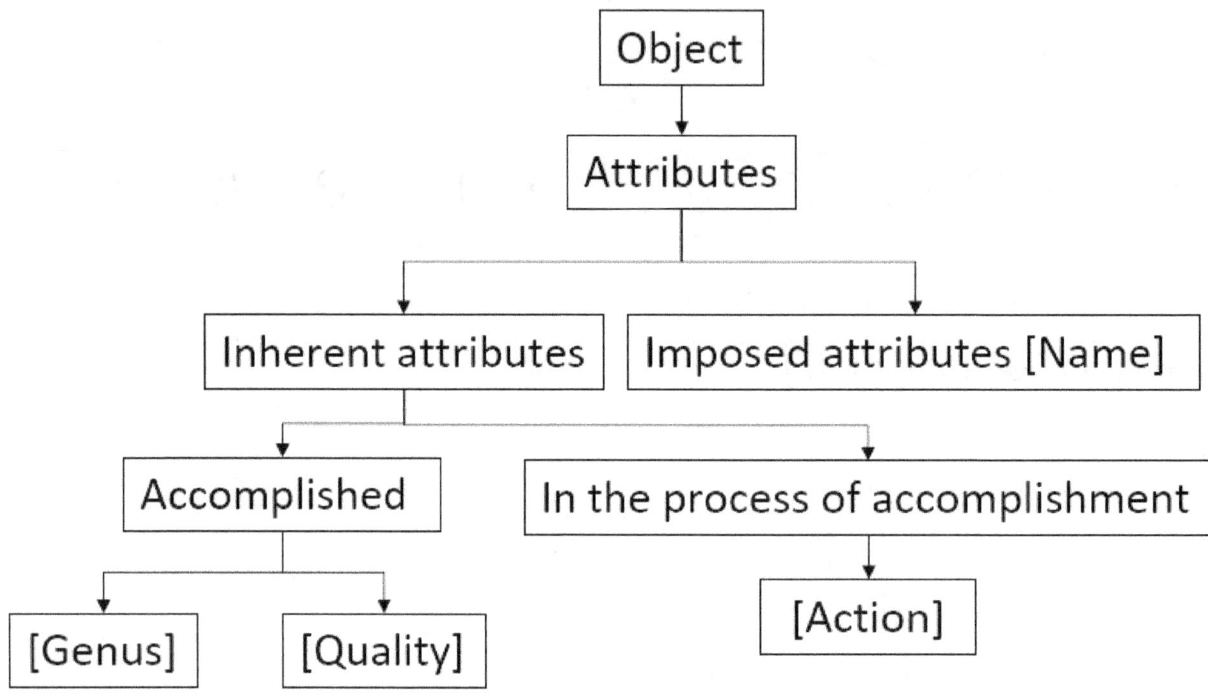

Figure 1. A thing and its attributes

Word meaning, where word indicates a thing according to the two schools of thought in India is as follows.
   1. Jatipaksa claim that the word refers to the class to which the individual/thing belongs.
   2. Vyaktipaksa claim that the word refers to the individual, who is a finished thing or an instance of a class.

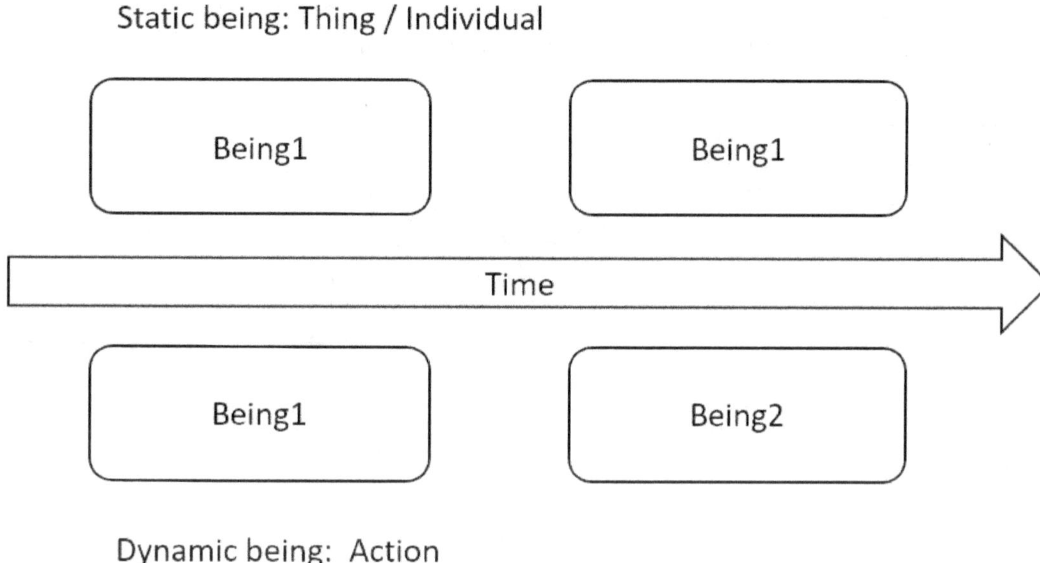

Figure 2. Static Being: Individual and Dynamic Being: Action

Being or Existence is of two types [1]:
    1. Static
    2. Dynamic

A static existence is a time-invariant existence which does not change with time. When there is a temporal sequence in existence it is dynamic or time-variant and action is said to take place [1]. The two types of being are depicted in Figure 2.

So Being which does not change with time is called substance, individual, "*dravya*" or the thing. It belongs to a jati which is the characteristics persistent across all changes. In Figure 3. we see that the Genus is the Being which is constant across a change and the Quality is the being which changes. So a thing is a collection of being which is constant and being which changes but at a given instance of time, it is the complete being of the thing at that instant of time.

Thing: Type of Being (Static)
Action: Type of Being (Dynamic)
Genus: Type of being which is persistent across change or something that continues to exist.

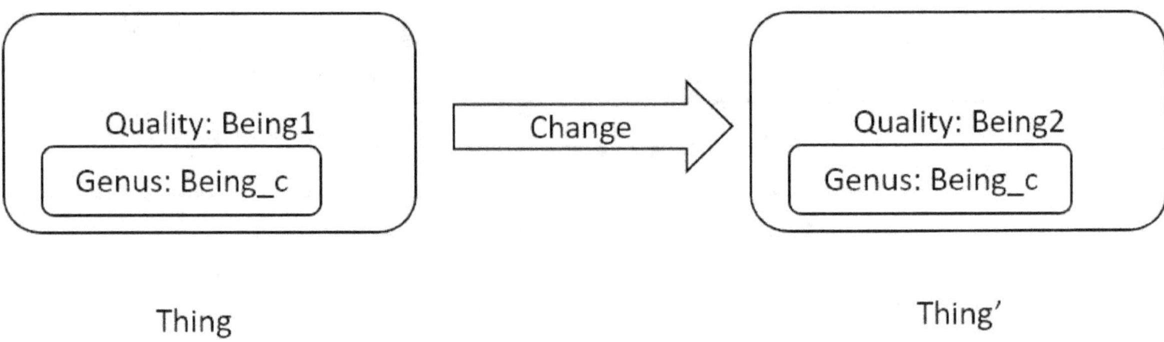

Figure 3. A Thing Changing

So a word can mean either the thing which it signifies or the genus the thing has, according to the schools of thought as shown in Figure 4.

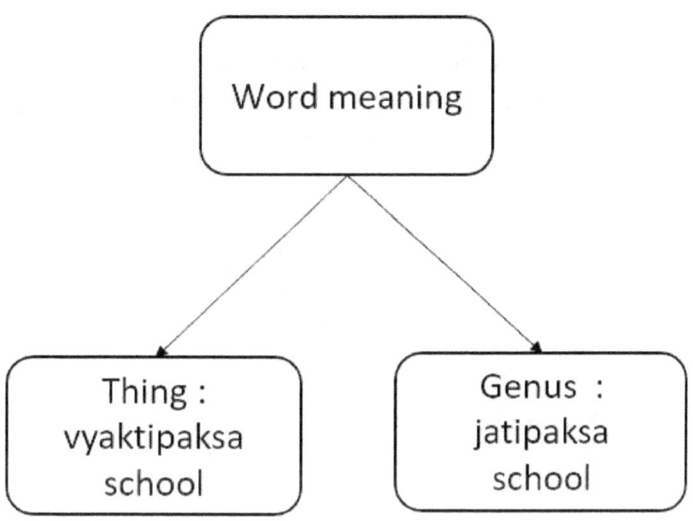

Figure 4. The two schools of thought and word meaning

So Being can be classified from our discussion as shown in Figure 5.

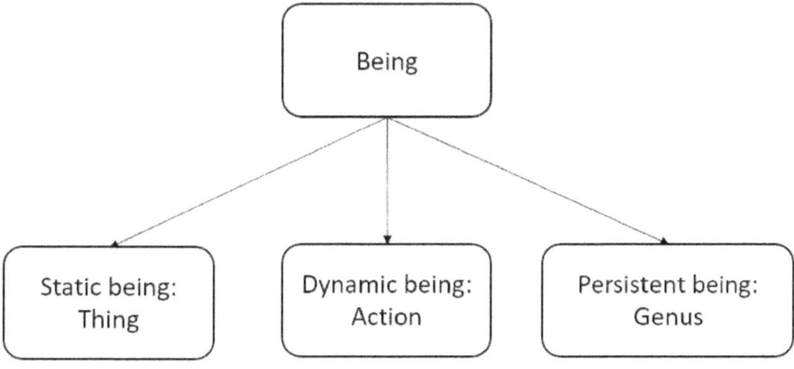

Figure 5. Types of Being / Existence

The implications for word meaning are shown in Figure 6. If a word refers to the genus as its meaning, the highest genus or Universal being Existence is what the word indicates. And a thing has been also shown to imply a static being or being itself. So whatever be the word meaning (Thing or Genus) it would always indicate a type of Being/Existence.

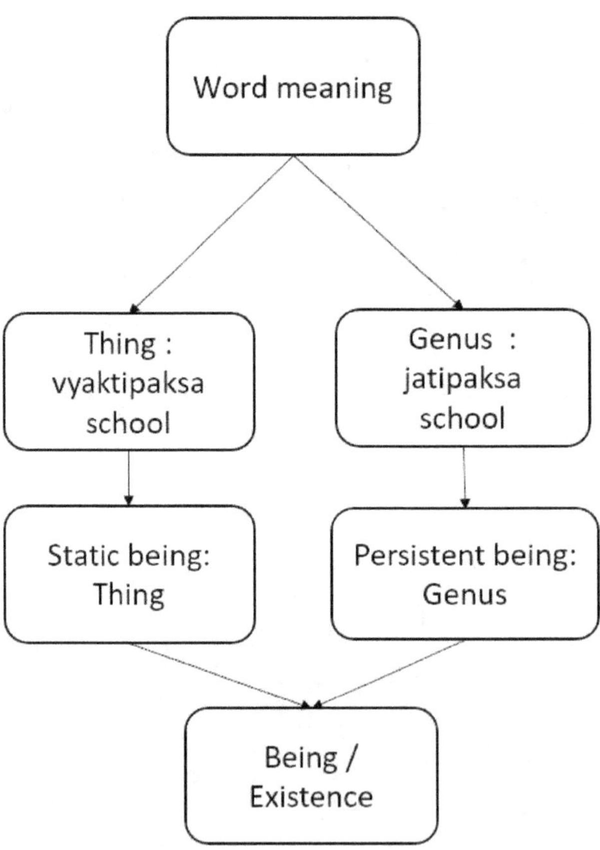

Figure 6. The two schools of thought on Word Meaning and Existence

So it has been proved that all words (which indicate a thing) indicate a type of "Existence" or "Being". Which implies they indicate 'Existence'.

Significance of all words conveying "Existence/Being":

- This existence or being is called *"satta"* in Sanskrit. Hence all words convey objects which are manifestations of *"satta"* ("Existence/Being"). [2]
- All objects are a type of "existence" or "being". They belong to the "absolute existence" genus.
- Existence inheres in all objects as even a non-existent object exists in the mind [3]. Absolute existence pervades all objects and also all objects are a type of it. It is everything and everywhere as well.
- Existence is one only. It is a common characteristic of all entities. It is all-pervading and filling all space, very large in its extent, and it is called *"Brahman"*.
- That which fills, that which swells, that which expands, that which is everywhere and in all things—That is the completeness, the fullness of Reality; and that is called *"Brahman"* in the Sanskrit language [4]. All existent things are nothing but a type of existence itself or pure existence hence everything is a type of existence. It sustains or upholds everything as without existence an object cannot be.

"Existence/Being" in ancient Indian religion:

- According to Hinduism the *'satta'* (existence) in each and all is God. In Hinduism, it is referred to as Paramatma or the supreme spirit [5].
- The all-pervading *'satta'* ("Being") is referred to as *Vishnu* in the *Puranas* (as it is all-pervading). *Vishnu* is the supreme god of Hinduism [6].
- "In the beginning, this [universe] was "Being/Existence" (Sat) alone, one only without a second". He desired, 'I shall become many and be born. He performed Tapas (austerities). having performed Tapas, He created all this (whatever we perceive). Having created it, He entered into it. Having entered it, He became the manifest and the unmanifest, the defined and undefined, the housed and the houseless, knowledge and ignorance, truth and falsehood, and all this whatsoever that exists. Therefore, it is called Existence', according to a sacred Hindu book [7].
- Existence is all existent and non-existent entities. Ahimsa and why to follow it stems from the fact, that he who injures living creatures, injures *Vishnu*(or God): for *Vishnu*(Existence) is all things [8].

Notes

1. Karl H Potter, Encyclopedia of Indian philosophies, Volume V Pg 108.
2. Vakyapadiyam of Bhartrhari, III.1.35 commentary.
3. Vakyapadiyam of Bhartrhari, III.1.34 commentary.
4. https://www.swami-krishnananda.org/upanishad/upan_06.html.
5. Vishnu Purana 6.4.37
6. Vishnu Purana 6.4.37
7. Taittiriya Upanishad – Brahmananda Valli – 2-6.
8. Vishnu Purana 1.2.10-13

CHAPTER THREE

The Existence of God and His Greatness with Scientific Proof

According to the Oxford English dictionary, God is (in monotheistic religions) the one Supreme Being, the creator and ruler of the universe. And from our previous chapters, we understood that everything is a type of "being" and the highest being is an "Absolute being" or "Absolute Existence". So the supreme being or the highest being is "Absolute Being" or "Absolute Existence".

In a softer sense, Merriam Webster dictionary defines God as an "object worshipped as divine". Here too as discussed before every "object" is a type of existence or it is existence itself. So when we worship an object we worship a type of existence and existence itself.

So in all scenarios, God is a type of existence and existence itself is worshipped or adored.
So God is one and God is "Absolute Existence" or "Absolute Being". Prominent schools of Indian Philosophy like the Vedanta also go by this definition of God and call him Brahman.

God according to prominent schools of Indian philosophy is Brahman, which means the which is "everywhere and is in everything". It is known as "Absolute Existence"[1] and "Absolute Existence" is worshipped as Brahman, i.e., God or Vishnu or Paramatma. Brahman is Shiva [2] or is Mother Goddess (Devi) [2]. The different names are aspects/manifestations of the Brahman.

We have already seen in the previous chapters that Brahman is "sat" or "satta matra", "the absolute existence". Existence has the following properties:

The generic part of everything (unchanging part) is the existence or a manifestation of existence.

1. It is the highest universal.
2. It is everywhere and in everything. This universal is in everything, hence things exist. Everything is a type of being and existence; hence it is everywhere.
3. "Brahman is the unchanging among changing things" which is the genus, hence that is Brahman/Existence.
4. So existence inheres in all objects as even a non-existent object exists in the mind. Absolute existence pervades all objects and also all objects are a type of it.
5. Existence is one only. It is a common characteristic of all entities. It is all-pervading and filling all space; very large in its extent, it is called Brahman, Vishnu.
6. It sustains or upholds everything as without existence an object cannot be.
7. It has no pain in it, hence it is joy or pleasure. Where pleasure is defined as the absence of pain.

Attributes of God ("Absolute Existence") which is manifested as everything movable and un-movable in the universe (provable) are:

1. He is the manifest and unmanifest.
2. He is the existent and the non-existent.
3. He is the defined and undefined.
4. The housed and the houseless.
5. Knowledge and ignorance.
6. Truth and falsehood.
7. And all this whatsoever exists [1].
8. Though Himself beyond any limiting concepts of time and space, the Eternal supports the cosmic manifestation in time and space as existence.
9. Though causing all the activity in the material world, He Himself does not act. As he(Absolute Existence) acts through his manifestations.
10. Though causing all the changes in the universe, He Himself is beyond any change. Absolute Existence is beyond change.
11. He remains unaffected by cosmic experiences.
12. He is the single binding unity behind all the diversity. Being self-existent, He does not need any support for existence.
13. Nothing can be without Him, and nothing can be beyond Him. Without existence, nothing would exist.
14. Knowing the Supreme Brahman one gains eternal bliss. His greatness is enlivening.
15. He is outside and inside of all beings. As Existence is everywhere.
16. He is non-moving and also the moving. As he is everything.

17. He is far away and yet too near as he is everything.
18. He is the one who is supporting all beings, destroying them; and creating them afresh.
19. Brahman is indestructible and the Supreme (higher than all else).
20. He is the formless and also the one with form. As he is everything.
21. Omnipresent, as he is everywhere
22. He is not restricted by space, time, and causality or by name and form, and is thus, infinite and omnipresent.
23. He is everything and everywhere.
24. The whole universe comprising all nature, and an infinite number of souls, is, as it were, the infinite body of God as all are him.
25. He is near and He is far, as he is both these objects. He is inside everything. He is outside everything, interpenetrating everything.
26. He sees through all eyes, hears through all ears, eats through all mouths, feels through all hearts, thinks through all minds, and reasons through all intellects, as he is everything.
27. With hands and feet everywhere, with eyes, heads and mouths everywhere, with ears everywhere, he encompasses everything in the world [4].
28. Formless Brahman assumes various forms for His own divine play (Lila) or sporting [4].
29. He is Agni (fire), Aditya (the sun), Vayu (air), and Chandramas (the moon). He is also the starry firmament. He is the Brahman (Hiranyagarbha), water, and is Prajapati (creator)[4].
30. He is woman, He is man, He is the youth. He is the maiden, too. He is the old man who totters along, leaning on the staff. He is born with his face turned everywhere.[4]
31. He is the dark blue fly. He is the green parrot with red eyes. He is the thunder-cloud, the seasons and the oceans. He is without beginning. He is the Infinite. He is from whom all the worlds are born.[4]
32. He is the bestower of blessings, who, though one, presides over the various aspects of Prakriti (nature), and in whom this universal dissolves, and in whom it appears in various forms [4].
33. He possesses countless heads. All heads, all eyes, all hands, all feet belong to the Lord. It is He who works through all hands, eats through all mouths, sees through all eyes, hears through all ears, walks through all feet and thinks through all minds [4].
34. It is He who, being one only, presides over every source of production and every form [4].
35. He is the internal Ruler of the universe.
36. His creation is constantly manifesting His powers by becoming new things.
37. His mind is the sum total of all the individual minds, and, therefore, it can be said that there is one agglomeration of mind and that belongs to God.
38. It is He, undoubtedly, who has become everything; but in some cases, there is a greater manifestation than in others.' 'God is everywhere. But there are different manifestations of His Power' like in Sun, moon, oceans etc. [4].

39. Being omnipresent in all locations allows Brahman-God to know everything that is occurring (omniscience) and to act everywhere with direct control over every part of the universe (omnipotence).
40. He is great because, as the sun he gives heat and light, as the moon light, as earth food and shelter, as the oceans and rivers water as your father, mother, brother and sister love and affection.
41. The enjoyer, the enjoyed and the enjoyment.
42. Appearing as Many due to the multiplicity of his powers [4].
43. He is grasped by the word 'OM' which is his name.

Some interesting knowledge which I have observed from various ancient Indian religious books:

Scriptures reveal that knowing or remembering his greatness (what you acknowledge is his greatness) causes love for him to arise in you (Bhakti) [5].

Scriptures reveal that worship is a form of adoring [6], as he is everything both animate and inanimate. You can love the animate or living and he will receive your love. As Maharaja Prahlada has mentioned in the Vishnu Purana "the whole universe is the manifestation of Vishnu. Search for the identity of Vishnu in all creatures. True worship of Vishnu consists in treating all equally" [7]. So love one, love all, as it is him only.

It is the Lord only, who is worshipped(adored) through actions, and from him, their results proceed [8].

According to a Hindu scripture [9] by always praising Vishnu, who is without beginning and end who is the Supreme Lord of all the worlds and who is the observer of the Universe, one gets beyond all grief. praising him continuously eradicates all suffering which includes diseases, poverty, pains etc. "It is further plain that praises, prostrations, etc., to Vishnu must be accompanied with perfect harmlessness to others". This is true as all are Vishnu and by hurting him the act of praising him is inconsequential as hurting and praising someone leads to pain for Vishnu and not pleasure and the goal of praising is to please him [9].

Adoring him is done through the process called "bhakti-yoga", where bhakti means love for love's sake and yoga means union or connecting with God. It has 9 forms [10] as given below:

1. Sravana [hearing of God's Lilas (divine play) and stories and songs about him],
2. Kirtana (singing of His glories),
3. Smarana (remembrance of His name, stories, qualities etc),
4. Padasevana (service of His feet),

5. Archana (worship of God),
6. Vandana (prostration to the Lord),
7. Dasya (cultivating the Emotion of a servant with God),
8. Sakhya (cultivation of the friend-emotion) and
9. Atmanivedana (complete surrender of the self)

The Supreme Person destroys all sins, when meditated upon, seen, sung, praised, worshipped, remembered or prostrated. Hence, He is Supreme purity [11]. Another manifestation of Vishnu, Krishna is identified as being anything which is the greatest or the most glorious of any class of objects [12].

Vishnu has two main forms (Sat and Vishwaroopam) and his other forms are incarnations or manifestations, his two main forms are [13]:

1. The first is "Sat" or absolute existence, it is attribute less, imperceptible and shapeless. "Sat" has an absence of pain hence it is pleasure or joy too.

2. The second is "Vishwaroopam" which is the universe (all moving and non-moving things) as we know it where he assumes all forms as a sort of divine play like gods, men, animals, non-living things and living things. It is the form of Vishnu which is manifested as every object of the universe. And meditating upon it i.e. understanding or contemplating upon it, destroys sins.

"Reciting/Saying the qualities of the Supreme Vishnu is the means to annihilate all sufferings in this material world" [14]. As it is a form of praising. And like the chanting of his name brings to mind his auspicious qualities which are purifying. And this activity is also the destroyer of all inauspiciousness and sins [15][16].

Some of his (Vishwaroopam) glorious qualities are:

1. Non-moving and moving.
2. Far and near.
3. Formless and formed.
4. Manifest and unmanifest.
5. Defined and undefined.
6. Housed and houseless.
7. Ascending and descending.
8. Rising and falling.

God's (objects of universe) qualities are wonderful and holy as you can see in the above list, enumerated are a list of opposites and in normal observation the presence of one opposite entails the absence of the other, but in God both the opposites are present at the same time which is unheard of and amazing to imagine. Multiplicity of powers in a single substratum lead to amazing combinations as well, for example he is a frame and he is a sapphire ring, so as a frame [is square, is made of steel] and as a sapphire ring [is circular, is made of sapphire]. Both being a single entity we can write Brahman [is square, is made of steel, is circular, is made of sapphire] here we can form wonderful combinations like he is a "square circle [is square, is circular]". Many such awe-inspiring combinations are possible that purify the intellect and heart. Combinations of glorious qualities are even more enlivening and all such glorious qualities are found in his *Krishna* manifestations (Sun, Moon, Sapphire etc [12]). Also as existence is all imaginary entities as well such combinations are also his manifestations only appearing in your mind.

"All sins are destroyed, and all good fortune is created by the Supreme Lord's qualities, activities and appearances, and words that describe these three things animate, beautify and purify the world" [16].

His names signify his qualities or actions, reminding us of his qualities which purify us and arouse love for him [17].

In ancient India, his names were considered medicines which can cure diseases when uttered [18]. Chanting the name aroused love for him and love for him is the embodiment of 'Ambrosia' or 'Amrit' [19].

Higher was the faith and love for him while chanting the name higher was its effect. "Faith is defined as the pleasing wishful contact with the object of pursuit" [20]. His names are infinite as his qualities are infinite. So you can also make up beautiful names for him (Vishwaroopam) and feel its purificatory effects. Vishwaroopam is easier to praise as you can see it with your own eyes. Sing his glories, celebrate him and feel the difference. But all this should be followed by harmlessness to all.

He is also the non-existent or the imaginary [21] so you can imagine something which gives you pleasure and praise it, as it is him only.

Notes:

1. Taittiriya Upanishad – Brahmananda Valli – 2-6.
2. Stella Kramrisch (1992). The Presence of Siva. Princeton University Press. p. 171. ISBN 0-691-01930-4.
3. David Kinsley (1988). Hindu Goddesses: Visions of the Divine Feminine in the Hindu Religious Tradition. University of California Press. pp. 136. ISBN 978-0-520-90883-3.

4. Swami Sivananda, Essence of the Svetasvatara Upanishad.

5. NELSON, LANCE E. "THE ONTOLOGY OF 'BHAKTI': DEVOTION AS 'PARAMAPURUṢĀRTHA' IN GAUḌĪYA VAIṢṆAVISM AND MADHUSŪDANA SARASVATĪ." Journal of Indian Philosophy, vol. 32, no. 4, 2004, pp. 345–92. JSTOR

6. Collins English Dictionary. Copyright © HarperCollins Publishers

7. Vishnu Purana 1.17.90

8. Swami Vireshwarananda, Brahma Sutras (Shankara Bhashya), Chapter III, Section II, Adhikarana VIII

9. R. ANANTHAKRISHNA SASTRY, THE VISHNU SAHASRANAMA WITH THE BHASHYA OF SRI SANKARACHARYA, Page 3, verse 4

10. Sri Swami Sivananda, BHAKTI YOGA, NAVA-VIDHA-BHAKTI section

11. R. ANANTHAKRISHNA SASTRY, THE VISHNU SAHASRANAMA WITH THE BHASHYA OF SRI SANKARACHARYA, Page 6, verse 10 commentary

12. Maharishi Ved Vyasa, Bhagavad Gita, Chapter 10, Verse 41

13. The Vishnu Purana a System of Hindu Mythology and Tradition Translated from the Original Sanskrit, and Illustrated by Notes Derived Chiefly from Other Puranas by the Late H.H. Wilson, Book VI, Chap VII

14. A. C. Bhaktivedanta Swami Prabhupada, Srimad Bhagavatam, SB 8.12.46

15. A. C. Bhaktivedanta Swami Prabhupada, Srimad Bhagavatam, SB 12.3.15

16. A. C. Bhaktivedanta Swami Prabhupada, Srimad Bhagavatam, SB 10.38.12

17. A. C. Bhaktivedanta Swami Prabhupada, Srimad Bhagavatam, SB 6.2.11

18. Sanatana Goswami, Hari Bhakti Vilasa, Eleventh Vilasa, Text 354

19. Narada Bhakti Sutra, Sutra 3

20. Baman Das Basu, The Sacred books of the Hindus, Page 38

21. Vakyapadiyam of Bhartrhari, III.1.34 commentary.